PANZER II

Horst Scheibert

Schiffer Military/Aviation History
Atglen, PA

Front cover artwork by Steve Ferguson, Colorado Springs, CO
Additional research by Russell Mueller.

AT THE GATES OF WARSAW
Illustrated here is a PzKpfw II Ausf. A, of the 4th Panzer Division scouting along a brick lane in the rain on the outskirts of Warsaw during the 1939 campaign in Poland. Unique for this era are the tri-pointed unit symbol on the forward armor plate, the black and white "balkenkreuz" painted over with yellow to obscure the national marking to Polish gunners, and the post mounted 7.92mm machine gun added for the tank commander's use.

Photos:
Federal Archives, Koblenz
Podzun-Verlag Archives
H. Scheibert Archives
Squadron Signal Archives
Heinrich Jürgens

Translated from the German by Dr. Edward Force,
Central Connecticut State University

Printed in the United States of America
ISBN: 0-88740-674-2

This title was originally published under the title, *Panzer II,*
by Podzun-Pallas Verlag.

We are interested in hearing from authors
with book ideas on related topics.

Published by Schiffer Publishing Ltd.
77 Lower Valley Road
Atglen, PA 19310
Please write for a free catalog.
This book may be purchased from the publisher.
Please include $2.95 postage.
Try your bookstore first.

Panzer II c, 6th Panzer Division (Panzer Unit 65), winter 1939-40. The c version, the first to be produced in large numbers, is recognizable by its rounded bow. All later types have flat bow plates.

Panzerkampfwagen II (Sd. Kfz. 121)

The development contract for the Panzerkampfwagen II was given to the firms of Henschel, Krupp and MAN in 1934. The final bonus was gained by MAN, and in 1935 the first tanks could already be sent to the troops under the designation of Panzerkampfwagen II (2 cm), Sd. Kfz. 121. The first two series (a and b) each included just a few tanks; they were in effect troop tests, so that only the third (c) series, with its familiar running gear of five large disc-type road wheels, quarter-elliptic springs and rounded bow was produced in large numbers. In addition to MAN, four other firms took part in its production by building it under license. The further versions, A to C and F, differed only in details, while all essentially had the flat bow plate unlike the rounded type of the earlier versions. The Ausf. C, and the F even more, also stood out from all the earlier types by having reinforced armor. These versions were built from 1940 almost to the end of the war.

In addition to these, there was one more development by Daimler-Benz. This differed from the above in having a more powerful engine, a different gearbox and Christie running gear (without return rollers), and by torsion-bar suspension. Its steering was also somewhat different, while its body, turret and armament conformed to those of Ausf. A, B, C and F. As a "fast combat wagon", it was intended for the "light divisions." Its various versions were designated D, E, G and I. But in relation to the MAN Panzer II, fewer vehicles were built.

The chassis of both types were also used during the course of the war as weapons carriers for antitank guns (Marder) and howitzers (Wespe), as well as for other purposes. There were also a flamethrowing version and an assault tank (Grille) with a 15-cm gun.

In all, there were the following Panzerkampfwagen II types:

121	Pz.Kampfwagen II (2 cm) Ausf. a1, a2, a3, b)
121	Pz.Kampfwagen II (2 cm) Ausf. c)
121	Pz.Kampfwagen II (2 cm) Ausf. A, B, C)
121	Pz.Kampfwagen II (2 cm) Ausf. F)
132	7.62 cm Pak(r) on Gw. II (F) "Marder II"
131	7.5 cm Pak 40/2 on Gw. II "Marder II"
124	10.5 cm le. FH 18.2 on Gw. II "Wespe"
	15 cm s.IG. 33 on Sf. II "Sturmpanzer II"
121	Pz.Kampfwagen II (2 cm) Ausf. D, E
122	Pz.Kampfwagen II (Flame)
133	7.62 Pak(r) on Gw. II (D, E) "Marder II"
	Pz.Kampfwagen II (2 cm) Ausf. G
	Pz.Kampfwagen II (2 cm) Ausf. J
123	Pz.Kampfwagen II (2 cm) Ausf. L, or
	Pz.Spähwagen II "Luchs"
	Ammunition tractor

A Panzer II c surmounting an obstacle in a demonstration in 1937. The weapons have been removed.

Panzerkampfwagen II

Ausf. a, b and c

There were only a very few of the a and b series made. Some were tested in the Spanish Civil War. In 1937 the Ausf. c, with its five big disc wheels and leaf springs, was already reaching the troops. Large numbers were produced, and the Panzer troops went into action in the campaigns against Poland, Norway, France and the Balkans equipped only – as far as the Panzerkampfwagen II was concerned –with the Ausf. c. However, tanks of this type would also later be found in Russia and North Africa. It differed from the later types – designated by capital letters – by its round bow (as opposed to the angular type beginning with Ausf. A).

A Panzer II Ausf. b of the 12th Panzer Division, with the swastika flag on the rear for aerial identification. Note the small road wheels, the chains on the bow for additional protection, the fog-cartridge launcher at the rear, and the trailer with a barrel of gasoline.

The technical data of Panzerkampfwagen II Ausf. c were:

Pz.Kampfwagen II Ausf. c, Sd. Kfz. 121
(MAN, Henschel, Famo) 1937

Engine	Maybach HL 62 TRM carbureted
Cylinders	6
Bore x stroke	105 x 120 mm
Displacement	6191 cc
Power	140 HP at 2600 rpm
Torque	41.5 mkg
Compression	1 : 6.5
Carburetor	1 Solex 40 IFF II
Valves	dropped, camshaft in cylinder head, gear-driven
Main bearings	8 interchangeable
Cooling	water, by pump
Battery	12 Volt, 105 Ah
Generator	600 Watt
Transmission	Rear engine, drive to track running gear, two-plate dry clutch, central shift
Gearbox	ZF Aphon SSG 45, 6 forward, 1 reverse, 2nd to 6th gears synchro nized
Chassis & body	Self-supporting armored hull, armored body with swiveling turret
Running gear	2 tracks, 108 links each (91 mm division), drive wheel in front, leading wheel at rear, 5 medium-sized road wheels in line, 4 return rollers, 5 quarter-elliptic springs
Steering & brakes	Planetary steering, mechanical steering brakes activated by two steering levers

A scene from Wehrmacht maneuvers in Mecklenburg in 1936. In the foreground, a company of Panzer Regiment I singing; in the background a Panzer II b. The significance of the red and white stripe around the turret of this tank is unknown.

General Information	
Track length	2400 mm
Track	1880 mm
Track width	300 mm
Overall dimensions	4750 x 2140 x 2000 mm
Ground clearance	340 mm
Wading ability	800 mm
Turning circle	4.8 meters
Fighting weight	8900 kg
Top speed	40 kph
Fuel consumption	road 110, off-road 170 liters/100 km
Fuel capacity	170 liters (2 tanks)
Range	road 150, off-road 100 km
Crew	3 men
Armor plate	Front 30 mm, sides & rear 14.5 mm
Armament	2 cm KwK 30 + one MG 34

Two views of a Panzer II b. The cross in the lower photo shows that this type of tank was still used in the campaigns against France and, in a few cases, Russia (1941).

Upper left: A Panzer II c of Panzer Unit 65 (1st Light Division), loaded on a low loader. The weapons have been removed. The tanks of the 1st to 3rd Light Divisions (Panzer Units 65-67) were all transported by truck and low-loader trailer, so they could be moved faster over great distances without wear and tear. But by 1938 these long and heavy road trains were eliminated because they were cumbersome.

Above: A Panzer II c (right) and Panzer IV of Panzer Regiment 2 (Eisenach) at the Ohrdruf training facility in Thuringia in 1936.

Left: Panzer Regiment 11 marches into the newly gained Sudeten area in 1938. Here a Panzer II c is greeted by the people of the city of Mies.

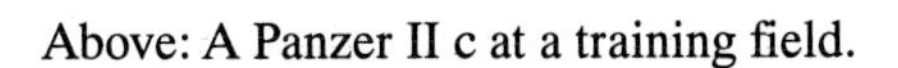

Above: A Panzer II c at a training field.

Upper right: Panzer I and II c of Panzer Unit 65 being loaded onto trucks and trailers at the Sennelager army base.

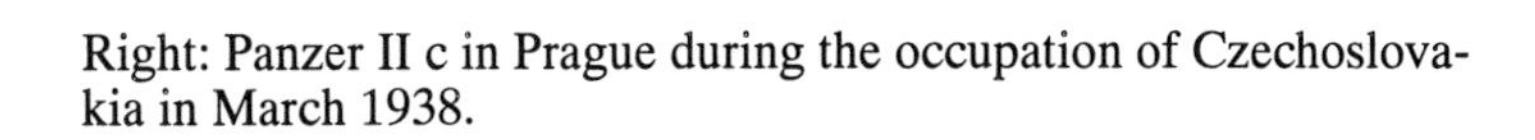

Right: Panzer II c in Prague during the occupation of Czechoslovakia in March 1938.

WH
131394

These two pages show photos from the 1939 Polish campaign. All the tanks are Panzer II c. The white crosses on the turrets and some of the hulls are typical of this campaign. Since they offered very good targets for Polish tanks and antitank guns, they were removed or painted over during the campaign.

On the left page (lower left) and on this page are photos of the final assault on Warsaw. At left is a Panzer II stuck fording a river near Radom; below is a broken down Panzer II.

The pennants seen at the lower left – usually made of sheet metal – usually pointed to command posts. In this case, the stripe in the middle is black and the rest of the surface is the color of the applicable service arm. The Roman numeral I indicates the first battalion. In the sense of the "combat of united weapons", these are probably the pennants of a rifle battalion (called Panzer Grenadiers as of 1942) and a Panzer unit.

Additional photos of Panzer II tanks in action in Poland in 1939. At left and below are scenes from Warsaw – later the grenadiers avoided the tanks as they drew enemy fire and thus did not offer much protection.

Below: A Panzer II c tows a broken-down six-wheel scout car.

On the right page are photos of parades in Warsaw (the two outside photos) and in Germany (Panzer Regiment 11 in Paderborn). The upper picture shows a Panzer II c on parade after the French campaign.

Parade in Warsaw (1939)

Panzer II c: rear view

Panzer II c: front view

NORWAY:
April 9, 1940
Operation
“WESERUBUNG”

A Panzer II c and a group of infantrymen advance on a road north of Oslo. Because of the very hilly terrain, tanks were used almost exclusively on roads in Norway. The Panzer II c is easy to recognize by the rounded bow plate and the lack of a commander’s cupola.

Left: This Panzer II is advancing along a railroad line. Two infantrymen are lying on it, in the cover of its turret. They jumped off when it moved ahead more slowly, as was usually the case when contact with the enemy was made. In addition, partisan activity began quite early in Norway, causing a very close relationship between tanks and grenadiers.

Panzerkampfwagen II

Ausf. A, B and C

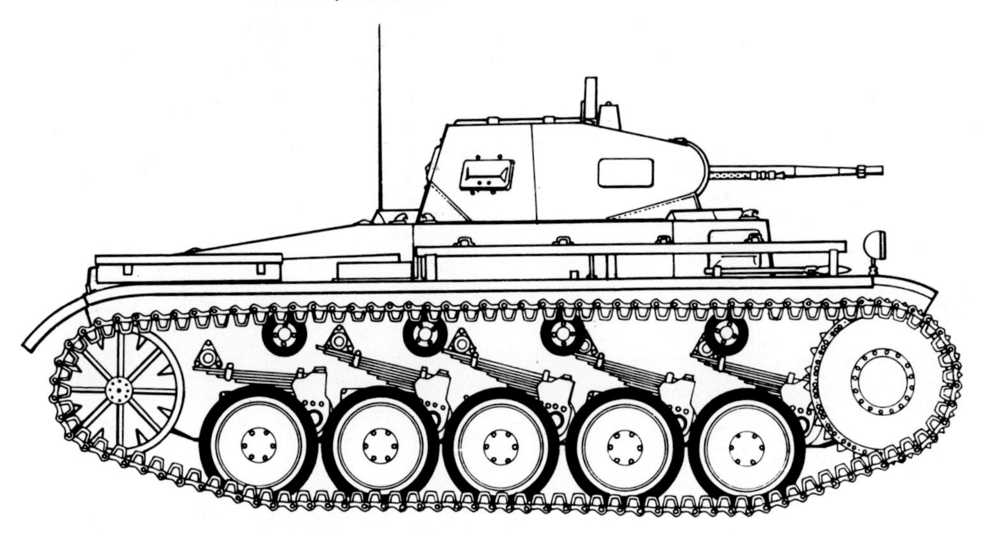

These versions differed from those previous not only in invisible details (see technical data), but also by their angled bow plates, which were given a certain reinforcement in Ausf. C, as well as by the commander's cupola on the turret, beginning with Ausf. B.

These Panzer II tanks were made in particularly great numbers and typified the Russian campaign. Many photos from this period thus show these versions of them.

The later Ausf. F. G and I differed externally with its smooth conical leading wheel, a somewhat higher commander's cupola, and from Ausf. G on, a sheet-metal "Rommel Box" on the turret.

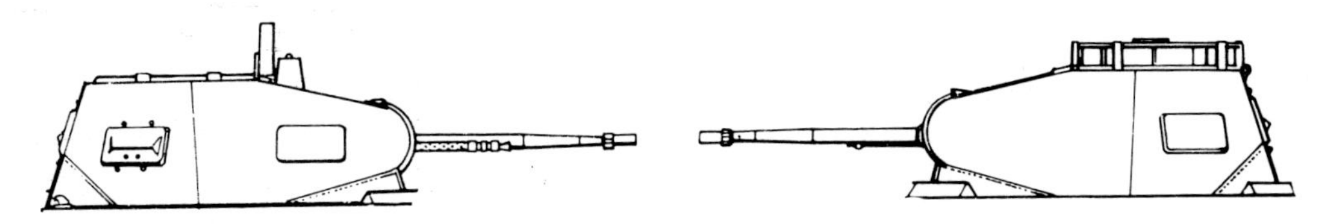

Turret of Panzer II
to Ausf. A

Turret of Panzer II
from Ausf. F on

Technical Data

Pz. Kampfwagen II
Ausf. A, B, C
Sd.Kfz. 121
(MAN, Henschel, Famo)
1938-1940

Engine	Maybach HL 62 TRM carbureted
Cylinders	6
Bore x stroke	105 x 120 mm
Displacement	6191 cc
Power	140 HP at 2600 rpm
Torque	41.5 mkg
Compression	1 : 6.5
Carburetor	one Solex 40 IFF II
Valves	Dropped, camshaft in cylinder head, gear-driven
Main bearings	8 interchangeable
Cooling	water, by pump
Battery	12 Volt 105 Ah
Generator	600 Watt
Transmission	Rear engine, drive to running gear, two-plate dry clutch, central shift
Gearbox	ZF Aphon SSG 46, 6 forward speeds, 1 reverse, 2nd to 6th gears synchronized
Chassis & body	Self-supporting armored hull, armored superstructure with swiveling turret
Running gear	Two tracks of 108 links each (91 mm intervals), drive wheel in front, leading wheel in back, 5 medium road wheels, 4 return rollers, 5 quarter-elliptic springs
Steering & braking	Planetary steering gear + me chanical steering brakes activated by two levers

In these photos of the Panzer II Ausf. A the turret is somewhat more sharply angled on the sides, and the entrance hatch on the top is square and made of two sections. The small projection at the front of the turret was for the optics.

Further technical data of Ausf. A, B and C:

General Information:

Track length	2400 mm
Track	1880 mm
Track width	300 mm
Overall dimensions	4810 x 2280 x 2020 mm
Ground clearance	340 mm
Wading ability	925 mm
Turning circle	4.8 meters
Fighting weight	9500 kg
Top speed	40 kph
Fuel consumption	Road 110, off-road 170 liters/ 100 km
Fuel capacity	170 liters (2 tanks)
Range	Road 150, off-road 100 km
Crew	3 men
Armor plate	Front 30, sides & rear 14.5 mm
Armament	2 cm KwK 30 + one MG 34

Black Panzer cap and headphones.

Before the French campaign began (May 1940), radio command, close cooperation between grenadiers and tanks, long marches with the appropriate supplying, and cooperation with the Luftwaffe were practiced in numerous drills. This Panzer II c of the 6th Panzer Division is seen at the Wahn training facility near Cologne.

The Campaign in France

There were Panzer II tanks in all the armored units during the French campaign, particularly in their reconnaissance platoons. Panzer II Ausf. A tanks (above and lower left) were more prevalent there than in the Polish campaign.

The II c, though, was still strongly represented (upper left and below). Close cooperation between infantry and tanks was always an important aspect during operations.

Right: Supplying during a pause in combat. At the same time, the empty double-drum magazines of machine-gun ammunition were exchanged for full ones. The empty ones are easy to see here on the track covers.

A Panzer II A of the 4th Panzer Division. The division's emblem is visible under the German cross on the turret. This is the third tank in the First Platoon of the 8th Company (813).

The commander wears the usual green coat of the army man over his black field uniform. There were never black overcoats in the army.

Lower right: In the second part of the French campaign, after grenadiers had attacked across the Aisne and, after hard fighting, broken through the Weygand Line, engineers erected bridges over which the tanks followed, in order to pursue the retreating enemy and thrust far ahead. The rolled-up corduroy road section on the rear of the Panzer II A was for use in swampy areas.

Left: Panzer II A in a Flemish city. Below: Panzer II A as platoon leader's tank of the reconnaissance platoon of the 1st Company, Panzer Regiment 11 (6th Panzer Division). The driver (Obergefreiter Kühne) has just been decorated with the Iron Cross, Second Class. In the visible form shown here, it was worn only on the day of awarding.

Above: A makeshift crossing of the Rhein-Marne Canal.

Below: Panzer II A of Panzer Regiment 2 in the victory parade in Eisenach. The oak leaf, an additional symbol of the 1st Panzer Division, is readily visible on the turret.

A Panzer II c of the 1st Panzer Division (note the official division symbol near the driver's lookout) forcing a river crossing. Engineers help to keep the torn-up roadway somewhat in order. The machine gun and 2 cm tank gun (BMK) have makeshift muzzle protectors of white cloth.

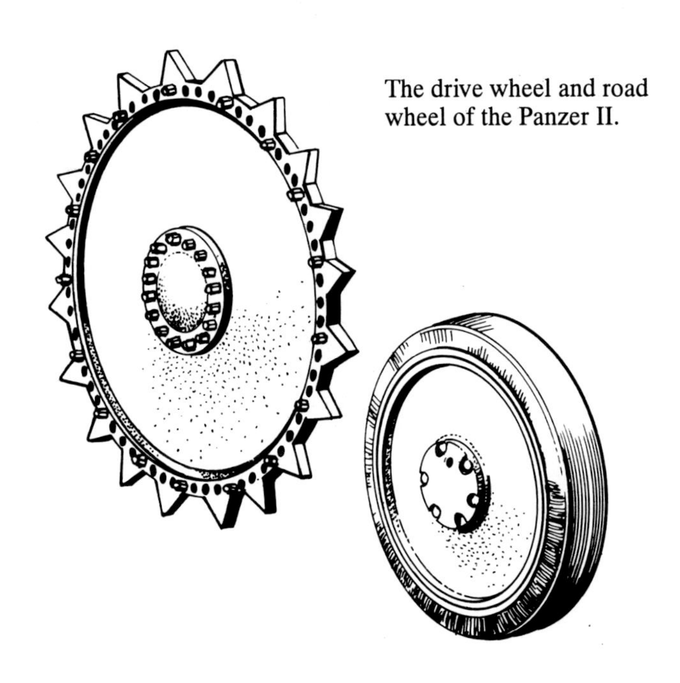

The drive wheel and road wheel of the Panzer II.

A Panzer II c of the 4th Panzer Division. The division emblem is easy to see near the driver's lookout and beside the German cross.

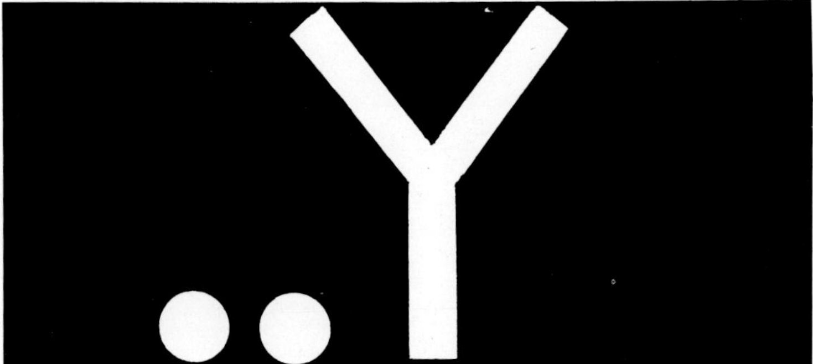

A Panzer II c of the 3rd or 6th Panzer Division crossing a log bridge. The division symbol, recognizable next to the driver's lookout and also as a sketch above, was used by the 6th Panzer Division until the end of 1940 but by the 3rd Division as of 1941. Since this picture was taken in the winter of 1940-41, it cannot be said for sure which division this tank belonged to.

A Panzer II c of the 4th Panzer Division drilling during the winter of 1940-41. It stands in firing position on a slope. Its division symbol is easy to see beside the back lookout of the hull and on the turret. Like all Panzer division symbols, this one was painted in yellow. One anomaly is the German cross on the turret, since one can also be seen in the usual position.

Likewise tanks of the 4th Panzer Division; in the foreground a Panzer II c, behind it a Panzer I. The commanders wear field-gray Army coats over their black uniforms. There were no black coats, but officers of the Panzer troops were also allowed to wear anthracite-colored leather coats.

Meeting in Africa; here a Panzer II B or C with a (long) SPW. The emblem of the Afrika-Korps (a palm tree with a swastika over the trunk) can be seen at right next to the German cross on the SPW.

Leading wheel of the Panzer II, Ausf. A, B and C.

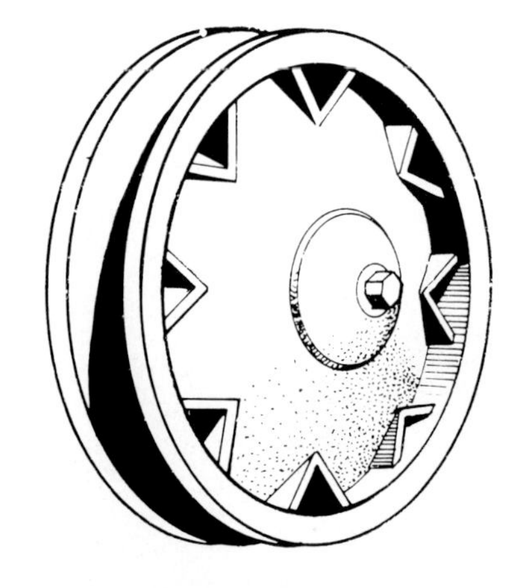

Leading wheel of the Panzer II, Ausf. F.

Two Panzer II B or C watch over the construction of a position in the desert. By the fires – presumably destroyed vehicles – one can tell that a battle or air raid took place shortly before.

A Panzer II Ausf. B of Panzer Regiment 5 of the 3rd Panzer Division – shipped on a freighter – on its way from Italy to Africa. Parts of this regiment had the first tanks of the German Afrika-Korps, which reached North Africa in the spring of 1941. Note the large numbers on the turret, plus the division symbol of the 4th Panzer Division formed with dashes instead of dots (see page 24); both forms were used.

Lower right: Panzer II of Ausf. B or C and F (left). The rear leading wheels indicate the various types. It cannot be told what unit they belonged to.

Panzer II F while tuning the radio traffic. The steel bulge above the driver's lookout (also seen in the upper right photo) was to prevent shots from reaching the turret ring (between turret and hull) and jamming or displacing it.

Panzerkampfwagen II

Ausf. D and E

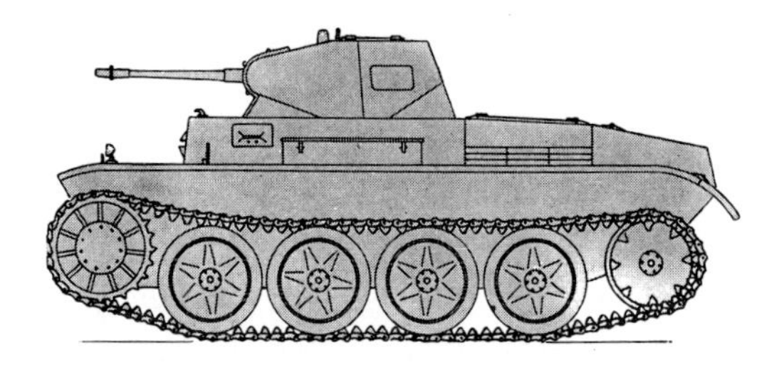

The D and E Ausf., recognizable by their four large road wheels (Christie running gear) with torsion-bar suspension, was a development of the Daimler-Benz works at Berlin-Marienfelde in 1938. It was intended as a "fast combat vehicle" to be used by the "light divisions." Along with the typical road wheels already mentioned, it also differed in having a stronger engine (180 instead of 145 HP), a Variorex pre-selector gearbox and clutch steering. The larger wheels certainly increased its speed, but in rough country they made greater demands on the running gear. Otherwise (body, turret, armament), it did not differ from the Ausf. C, though it did have somewhat thicker armor. The differences between the D and E types were minor and not externally visible.

After the "light divisions" had been reequipped with Panzer 35 (t) and 38 (t), the Panzer II D and E were used wherever the other Panzer II were used. In 1941 production was halted; about 250 had been built.

Panzer II D or E tanks in a loading drill (1939), being loaded onto trucks or low loaders. The independent tank units belonging to the "light divisions" carried their tanks on trucks and low-loader trailers. These road trains, though, were so awkward that use of them stopped as early as 1938.

These photos show Panzer II D or E tanks. To the right and left behind the turret, attached to the track aprons, are the three launchers for smoke cartridges. They could be operated from the fighting compartment.

The meaning of the three semicircles on the turret is unknown; they may have been the symbol of a regiment or other unit. The steel helmets carried on the turret of the tank at lower right refer to the many head injuries of the Panzer soldiers during the war. They were worn when the crews had to leave the tanks because of "iron-bearing air."

Panzerkampfwagen II F

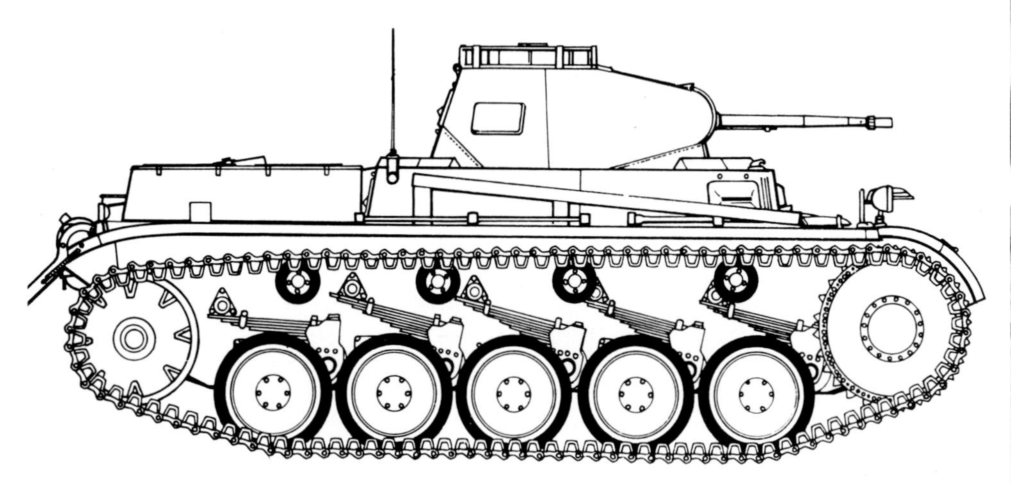

The Panzer II F first appeared at the end of 1940. The most important changes from the earlier versions were the strengthened armor plate and the conical leading wheel. The bow armor was increased to 36 mm, the armor of the driver's compartment to 31 mm, and the rest of the chassis armor to 20 mm. The turret armor remained the same as that of Ausf. A, B and C except for the front plate. This gave the tank a weight of over ten tons, which put much pressure on the engine, which still produced only 140 horsepower.

A new gun, the 20 mm Kampfwagenkanone (KwK) 38, was installed in place of the old type. It had the same rate of fire as the KwK 30, however the lengthening of the barrel increased the muzzle velocity of the shell and thus the effectiveness of the gun. The machine gun remained as a secondary weapon, but its ammunition was increased to 2550 rounds.

Although changes in the armor plate and the primary weapon had been made, this tank did not prove to be as good as the earlier series. It was inferior to all enemy tanks, because they were equipped with a stronger primary weapon. Thus these changes remained in production for only a few months, and during 1942 production of this tank version was halted. More and more chassis were rebuilt or modified into vehicles for special purposes. Yet it must be pointed out that in the years of the Wehrmacht's greatest victories (1939-1941), the Panzer II was used by a great many of the Panzer troops and had proved itself well in the hands of a capable crew.

A Panzer II F seen from the rear. The commander's cupola on the turret, with the periscopes that afforded a good panoramic view, can be seen easily.

A Panzer II F of a regimental staff. The R on the turret shows that. The 06 means that it belongs to the regiment's reconnaissance platoon. This tank was captured by the British during the African campaign and is to be seen today at the Royal Tank Museum in Bovington, England.

The reinforcements to the driver's area and front end of this Panzer II F are visible here. The conical leading wheel can also be seen clearly.

Panzerkampfwagen II F

(Sd. Kfz. 121)

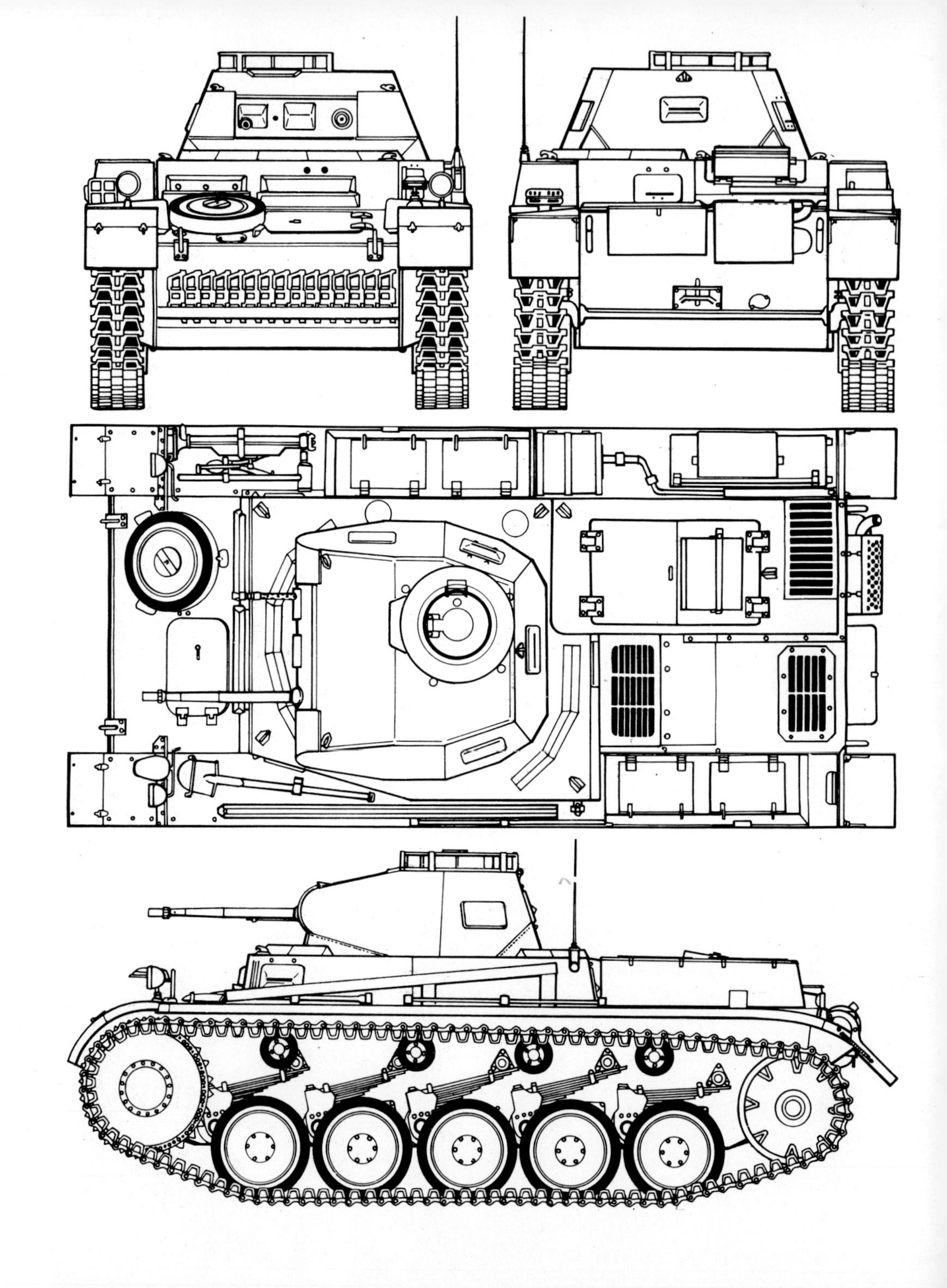

Technical Data

Manufacturers	MAN, Wegmann, Miag
Years built	1940-1943
Number built	625
Crew	3 men
Weight	9.5 tons
Length	4.81 meters
Width	2.28 meters
Height	2.02 meters
Engine	Maybach HL 62 TRM 6-cylinder, water-cooled
Power	140 HP at 2600 rpm
Transmission	5 forward speeds, 1 reverse
Steering	lever steering
Track width	300 mm
Ground clearance	340 mm
Armor plate	Front 30-35 mm, sides 14.5 mm, rear 20 mm
Armament	one 20 mm KWK 38 L/55, one MG 34
Ammunition	180 20 mm, 2550 7.92 mm rounds
Fuel capacity	170 liters total, in 2 tanks
Fuel consumption	road 110, off-road 170 liters/100 km
Range	road 150 km, off-road 100 km
Top speed	40 kph

The War in Russia

The last Ausf. c Panzer II were built in 1937, the last Ausf. A, B and C in 1940, the last Ausf. D and E in 1941 and the last F in 1942. All of them still saw service in Russia. Thus almost all types will be found in the following photos. As of 1942, more and more of their chassis were used only as weapon carriers.

On June 22, 1941 the Russian border was crossed in its entire length, from the Baltic to the Black Sea, and there was combat in the first hours. Each of the Panzer regiments and units had a reconnaissance platoon, usually of five Panzer II tanks. Sometimes there were also Panzer II tanks of any type in the combat companies. The letter K in the upper photo indicates membership in Panzer Group (Army) 1, which was commanded by Generaloberst von Kleist.

A Panzer II B column on the march in the central sector of the eastern front in the summer of 1941. The swastika flag on the engine cover identified the tank to the Luftwaffe. The crew's coats and blankets are bundled up and tied to the turret.

Lower left: Panzer II F – recognizable by the conical shape of the leading wheels – during a lull in combat on the steppes of southern Russia in 1942. On the inside of the opened turret hatches one can see the leather upholstering.

While the artillery fires on a target in the background, armored vehicles wait for the order to attack further. At left are two Schützenpanzerwagen (SPW), in the middle a long SPW with frame antenna, at right a Panzer III and in the foreground a Panzer II.

Upper left: Enemies meet. A Panzer II passes an abandoned Russian T-26 C tank. The latter is armed with a 4.5 cm tank gun. Both of them were already opponents in the Spanish Civil War.

In the foreground is a Panzer II C of the 4th Panzer Division (division symbol on front plate). In the background, mechanics work on a Panzer III with one end raised off the ground.

The wave rising to the driver's lookout shows the great wading ability of the Panzer II. Along with the rest of the crew, an infantryman sits on the tank. The tank's weapons are raised and fitted with muzzle protectors.

Panzer II C in the northern part of the eastern front in 1941.

Below: Grenadiers riding on Panzer II tanks on the eastern front in 1941. During lulls and after battles, tanks were often used as "transport vehicles." Right: a Panzer II F. Here the spare track links and road wheels, necessary as replacement parts and also used as extra armor, can be seen. The spare fuel canisters on the right track apron are also visible.

Above: A Panzer II belonging to an armored engineer company. The apparatus on the turret was used to ignite laid charges. At left the additional front armor, added to the turrets of the later Panzer II types, can be seen.

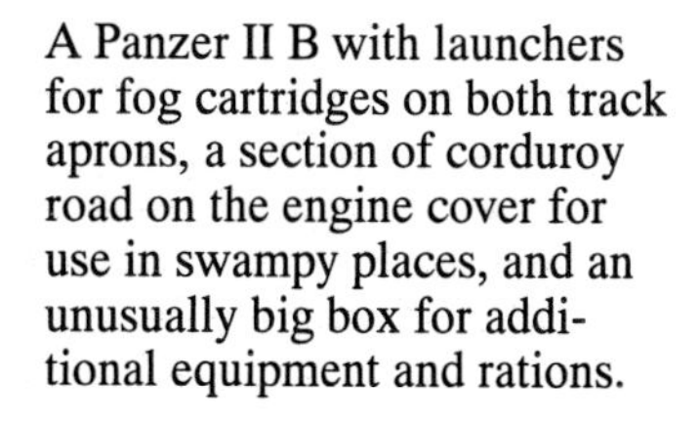

A Panzer II B with launchers for fog cartridges on both track aprons, a section of corduroy road on the engine cover for use in swampy places, and an unusually big box for additional equipment and rations.

A Panzer II F meets a “Stork.” This Fieseler Fi 156 “Storch” courier plane brought reports (which were dropped), transported liaison officers and commanders and could land almost anywhere.

Panzer II C in rough country on the northern front in the spring of 1942.

A Panzer II C of the 4th Panzer Division. It is the 6th vehicle of a unit (battalion) staff and thus probably belonged to the reconnaissance platoon.

A Panzer II B of the 7th Panzer Division. The division symbol can be seen near the left foot of the Leutnant riding on the tank. Eastern front, 1941.

A Panzer II C. This type was built in large numbers until 1940. The tools for the 2-centimeter BMK tank gun were kept in the box on the track apron.

The same tank as above from another angle. The BMK fired from magazines and required good maintenance.

This photo of a Panzer II C very clearly shows the additional armor plate riveted onto the front of the turret and ahead of the driver.

A Panzer II crosses a military bridge over the Akssay (Kalmuck Steppes). It has winter camouflage paint and is followed by a halftrack towing tractor. Panzer Regiment 11, 6th Panzer Division, December 14, 1942.

A Panzer II C in the winter of 1942-43. The symbol by the driver's side lookout is the tactical symbol of the Wehrmacht for "tank"; the 1 beside it means "first company."

A Panzer II C in action on the eastern front in 1943. On the track apron are launchers for fog cartridges, and on the sides are track links for added protection, plus a steel helmet on the turret and the always necessary bucket in back. The grenadier beside the tank carries an MG 42 machine gun.

A Panzer II B in a demonstration, in which both Army and Waffen-SS officers are taking part.

A Panzer II C of the 8th Panzer Division on a Russian road. The division symbol can be seen at left near the turret hatch. The Panzer II was a technically reliable tank, but in terms of armament, the 2 cm tank gun had its faults; all in all, the Panzer II was already obsolete by 1940.

A Panzer II C of the 4th Panzer Division (division symbol at right by the cross) after being towed out of a swamp. Number 106 means: 6th tank (probably reconnaissance platoon), 1st unit (= battalion), Panzer Regiment 35.

Panzer II standing ready for an attack on the steppes of southern Russia. The meaning of the three-part crosses is unknown; they were probably a troop recognition symbol.

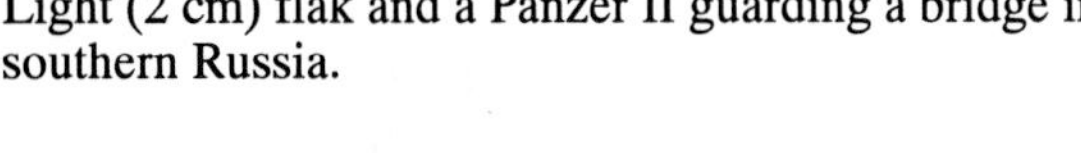

Light (2 cm) flak and a Panzer II guarding a bridge in southern Russia.

The crew of a Panzer II takes three prisoners on the steppes of southern Russia. The driver's side lookout is partly opened for better ventilation.

Out of the Panzer II D and E (developed by Daimler-Benz), there came, starting in 1942, the Ausf. G (lightly armored) and Ausf. I (heavily armored). Finally, the last was developed into the "Luchs" (Lynx) armored scout car, a type that proved itself well. It was intended to replace the halftrack scout car.

Naturally there were also special versions of the Panzer II. For example, about a hundred "Schnellkampfpanzer" of Ausf. D and E were rebuilt into flamethrowing tanks. Then captured Russian 7.62 cm antitank guns were mounted on Ausf. D and E chassis and thus turned into self-propelled antitank gun carriers. Later German 7.50 cm antitank guns mounted on the Panzer II A-C chassis became the Marder II Finally, Panzer II A-C chassis were rebuilt into weapon carriers for the 10.5 cm field howitzer (Wespe = Wasp), while small numbers were fitted with the 15 cm infantry gun as a "Sturmpanzer" (assault tank).

A Panzer II in a village in the Ukraine in 1944. In the background are two Sd. Kfz. 131 (Marder II) with the German 7.50 cm Pak gun on Ausf. A-C Panzer II chassis.

Also available from Schiffer Military History

OPERATION BARBAROSSA
OSTFRONT 1944
ROMMEL IN THE DESERT
PANZERKORPS GROßDEUTSCHLAND
THE HEAVY FLAK GUNS
THE WAFFEN SS
THE 1st PANZER DIVISION
THE GERMAN INFANTRY HANDBOOK